Mimi MARIAM KASHEGA

Bacteriological analysis of street foods

AF180339

Mimi MARIAM KASHEGA

Bacteriological analysis of street foods

in Kisangani's Kabondo commune (Malewa case)

ScienciaScripts

Imprint
Any brand names and product names mentioned in this book are subject to trademark, brand or patent protection and are trademarks or registered trademarks of their respective holders. The use of brand names, product names, common names, trade names, product descriptions etc. even without a particular marking in this work is in no way to be construed to mean that such names may be regarded as unrestricted in respect of trademark and brand protection legislation and could thus be used by anyone.

Cover image: www.ingimage.com

This book is a translation from the original published under ISBN 978-620-6-72989-1.

Publisher:
Sciencia Scripts
is a trademark of
Dodo Books Indian Ocean Ltd. and OmniScriptum S.R.L publishing group

120 High Road, East Finchley, London, N2 9ED, United Kingdom
Str. Armeneasca 28/1, office 1, Chisinau MD-2012, Republic of Moldova, Europe
Managing Directors: Ieva Konstantinova, Victoria Ursu
info@omniscriptum.com

Printed at: see last page
ISBN: 978-620-6-76292-8

DEDICATION

To our late mother

We dedicate this work

Mimi **MARIAM KASHEGA**

ACKNOWLEDGEMENTS

With all due respect, we give thanks to the Eternal God who has given us the strength and determination to transcend all moments of trial and hard work.

Our sincere thanks go to all the professors, supervisors and assistants at the Faculty of Science in general, and the Biotechnology Department in particular, for their educational, scientific and moral contributions throughout our time at the University.

Our most distinguished thanks go to Professors René OLEKO WOTO and Didy ONAUTSHU ODIMBA, respectively director and co-director of this work. Despite their many administrative, academic and scientific commitments, they agreed to supervise this work, which marks the end of our university career.

We would also like to thank Assistant Adelphine OMBA MIANGO, the supervisor of this work, for agreeing to direct this work and contributing both scientifically and materially to its completion, despite their multiple tasks.

We would also like to thank André TSHITENGE, technician at the Faculty of Science's microbiology and phytopathology laboratory, and assistants Martin ASSUMANI and Vincent MONGENGO for their technical and moral contributions to the completion of this thesis.

We would like to thank all the members of the KASHEGA family in general, and in particular: our biological father Gabriel K. for his love and involvement in my studies, our tutor Christian K. for believing in us throughout our academic career, and also our sister Naomie K. for her unconditional love and motivation in finalizing this work. Finally, we would like to thank Maman Fatouma, Maman Ruth, our grandmother Bi Maria, Pasteur Théophile, Auntie Angel and Papa Dave, for their prayers and above all for always being there for me.

To my fellow members of the audience who contributed in some way to the feasibility of this work, and to our friends and acquaintances for their support and encouragement

To one and all, we say a big thank you.

Mimi MARIAM KASHEGA

SUMMARY

BACTERIOLOGICAL ANALYSIS OF FOOD SOLD ON PUBLIC ROADS IN KISANGANI IN THE COMMUNE DE KABONDO (Case of Malewa)

The aim of this study was to assess the microbiological quality of a group of street meals sold at various restaurants (Malewa) located along the main road to Camp Ketele, in the Commune of Kabondo in Kisangani, in accordance with AFNOR standards. The meals analyzed consisted of rice, beans and pondu

Samples were taken at 3 different restaurant sites along the road to Camp Ketele.

The main germs tested were total aerobic mesophilic flora (TAMF), total coliforms (TC), *Staphylococcus aureus* and Salmonella sp.

The hygienic quality of rice, bean and poppy samples was assessed on the basis of microbiological criteria.

Susceptibility testing of strains to antibiotics commonly used in Kisangani was carried out.

The results found this study highlighted the presence of spoilage flora (FMAT), contamination indicators (total coliforms, in our case) and potentially pathogenic germs (S. aureus and Salmonella sp.) in all the samples analyzed, in our various target sites along the Camp Ketele road in the commune of Kabondo.

The bacterial load found was higher in the second restaurant than in the first and third

The hygienic quality of sampled foodstuffs is deteriorating at all target outlets. In line with quality standards, all the samples analyzed were classified as unsatisfactory, presenting a high risk of food poisoning for the health of the consuming population.

Potentially pathogenic strains *of S. aureus* and Salmonella sp are sensitive to Ciprofloxacin and resistant to all other antibiotics.

In fact, the lack of hygienic measures during the preservation and cooking of rice, beans and potted rice during preparation, the tools used, the poor preservation conditions (lack of cold), the various handling operations (cutting and selling) at these restaurants (Malewa) in the commune of Kabondo, the method of preservation or type of packaging of the food, and the healthiness of the restaurant environment itself, all explain the poor hygienic quality of the rice, bean and potted rice samples analyzed.

Key words: Hygienic quality, bacterial load, food, restaurant, street.

LIST OF ABBREVIATIONS

CT: total coliforms

S. aureus: *Staphylococcus aureus*

FMAT: Total aerobic mesophilic flora

EMB: Eosin methylene blue

MSA: Mannitol salt

T°: temperature

WHO: World Health Organization

S. paratyphi A: *Salmonella paratyphi* A

S. paratyphi B: *Salmonella paratyphi* B

MIU: indole urease mobility

TABLE OF CONTENTS

INTRODUCTION

1. Issues

The problems of undernourishment and malnutrition in developing countries, particularly in the Democratic Republic of Congo (DRC) and more specifically in the city of Kisangani, are of concern to many researchers and international organizations such as the WHO (Ngabu, 2005). Food-borne diseases are numerous and cause adverse effects on human health. These diseases are transmitted by vectors (flies, insects, etc.) that circulate freely from a contaminated area to an uncontaminated one; but also by lack of hygiene (WHO, 2003; Euloge et *al.*, 2012).

Kisangani is currently experiencing intense activity in the sale of ready-to-eat foods in various public places (markets, railway stations, streets, traffic circles, etc.). In many towns and cities in the DRC, a significant proportion of the urban population is consuming urban street foods (AVP).

Given that these foods are sometimes sold at affordable prices, less expensive and above all that they constitute a daily meal ready to be served or consumed immediately, several families are dedicated to selling ready-to-eat foods in various public places (Ambegebo., 2016).

However, AVP vendors are organizing themselves to set up small restaurants and stalls next to and in public places (markets, ports, stations, streets or traffic circles, etc.). To carry out their sales (fish, grilled meats, grilled caramel peanuts, omelettes, doughnuts, etc.); Without ignoring that these can be contaminated by various factors such as dust, flies, wind, buyers, etc. (Pene., 2020).

Food poisoning has been on the increase over the last twenty years. They can be the source of serious infections, which is why good food hygiene is so important. Food safety is an essential public health issue for all countries (Kavira., 2010).

This is how it is said that P.V.A.s constitute a major source of contamination, following their conservation during sale and handling of materials along the preparation process. These foods can also be the source of a number of human diseases, known as "hand" diseases: cholera, bacillary dysentery, amoebic dysentery, typhoid fever, salmonellosis, etc. (Makengo., 2002).

Of course, for a food product to be ready for consumption, it must not contain anything dangerous or toxic for consumers. Potential sources of contamination for these foods are: soil, water, air, humans, equipment, packaging materials (Mughole., 2012).

It's therefore important that hygiene standards for vendors are applied, to avoid the proliferation of microorganisms that are dangerous to human health.

Given that the vendors only take their own interests into account, and that they feel completely free because no control system is in place, we have decided to carry out this study based on the bacteriological analysis of food sold on public roads. With this in mind, we decided to carry out a study based on the bacteriological analysis of food sold on public roads. This is an important research question in the field of food hygiene and public health, and aims to assess the microbiological quality of food offered for consumption in these places, in order to identify possible health risks for consumers.

2. Assumptions

Since these foods are sold in public places, we thought that..:

- ✓ People who use PVAs are at risk of contracting diseases.
- ✓ The bacterial loads of these foods (rice, beans, etc.) do not comply with AFNOR standards;
- ✓ The environment is unsuitable for most of the sales outlets targeted in this study.

3. Objectives

3.1 General objective

Our aim is to verify the microbiological and hygienic quality of AVP during preparation, as well as the risks associated with the environment.

3.2 Specific objectives

The specific objectives of this work are as **follows**

- ✓ Isolate and enumerate FMAT, C.T, Salmonella and Staphylococcus in meals (rice - beans - beans) sold along the Camp Ketele road.
- ✓ Determine bacterial loads of FMAT, C.T, Salmonella and Staphylococcus in menus to assess microbiological quality;
- ✓ Appreciate the quality of the environment in which food is prepared and sold.

4. Interest

This work is of twofold interest

- **From a scientific point of view**: This investigation would make a real contribution to knowledge of the hygienic quality of the AVP sold along the Camp Ketele road.
- **From a hygiene and public health point of view**: This study will enable sellers to comply with basic hygiene standards, and consumers and the authorities in charge of hygiene and public health to be aware of the risk of contamination of AVP by FMAT, C.T, Salmonella and Staphylococcus.

5. Previous work

- **Ambengebo. 2016:** Microbiological quality of food sold on public roads (Case of "Liboke" sold at the Markets of Kisangani, DR Congo).
- **Kama, 2015**: Hygienic quality of food sold on the public highway during the day and at night (case of the 11th Tshopo market).
- **Lambewa; Mpiana. 2015:** Contribution à l'étude des aliments vendus sur la voie publique au marché diurne et nocturne à Mangobo (Cas de Limanga ya Nse).
- **Lokangila, 2013:** Hygienic quality of street foods in Kisangani - case of the IAT market (FMAT and Salmonella contamination of AVP).
- **Nzoto. 2016:** Contribution à l'étude de la qualité bactériologique des aliments vendus dans la voie publique de la ville de Kisangani (RDC): cas du manioc préparé (masele).
- **Swedi., 2022:** Microbiological quality of roasted and caramelized peanuts sold in the public streets of the city of Kisangani, more precisely in the vicinity of public markets.

6. Labor Subdivision

Apart from the introduction and suggestions, this work is divided into three chapters:

- The first deals with generalities,
- The second deals with materials and methodology,
- Third is presentation, interpretation of results and discussion.

CHAPTER ONE: GENERAL INFORMATION ON STREET FOODS (AVP)

1.1. Definition of concepts

✓ **Food**

It's a substance consumed in its natural state after cooking, capable of providing materials for growth and sustaining the body's services. All foods are made up of a small number of simple elements, including proteins, carbohydrates, lipids, water, minerals, vitamins and fibers (LAROUSSE MEDICAL., 2009).

Foods are usually classified into the following six groups:

- ✓ Group I foods: milk and dairy products

Provide animal proteins. Their fat and carbohydrate content varies. They are very rich in calcium, vitamins A, B2 and B12, but contain little iron and practically no vitamin C.

- ✓ Group II foods: meat, fish, eggs

They are high in protein. They provide iron and B vitamins.

- ✓ Group III foods: Fats: butter, margarine, cream, oils

They represent only the visible fraction of animal fat intake. They are rich in lipids, vitamin A (butter, cream) and vitamin E (margarine, certain oils).

- ✓ Group IV foods: cereals and derivatives, potatoes, dried vegetables

They are carbohydrate-rich energy foods, containing from 10% (cereals) to over 20% (pulses) protein.

- ✓ Group V foods: fruit and vegetables

They are rich in water, minerals and vitamins, with a carbohydrate content ranging from 5 to 20%. The cellulose they contain helps to facilitate intestinal function.

- ✓ Group VI foods: sugar and sweet products, honey, jam, chocolate, syrup, fruit juices, etc.

They form carbohydrates that can be rapidly used by the body (Larousse médical., 2009).

✓ **Street foods**

According to the FAO (1990), street foods are foods and beverages prepared ready for consumption and sold by street vendors or semi-stationary vendors in streets or other public places (market, port, railway station, etc.). These foods account for a significant proportion of consumption.

The informal food sector has been defined as "the sector producing ready-to-eat food and beverages, prepared and/or sold by vendors, especially in streets and similar public places" (FAO/WHO., 2007). At first glance, it may appear quite marginal when compared with the major sectors of agricultural production, the agri-food industry and the marketing of agricultural and food products.

For sellers, the sale of MVAs constitutes an important and regular source of income, as it is seen by less-educated individuals as an employment that could alleviate their needs in daily life, and alleviate the cost of unemployment in poor countries. On the other hand, consumers consider MVAs to be the cheapest and most accessible way of obtaining a nutritionally balanced meal away from home (Secke., 2007). Consumers are unaware of the hygienic conditions under which the food is prepared and preserved, and are sure to expose themselves to health risks.

In many countries, MVAs make a significant contribution to employment, income and household food security. They also help to meet the challenge of feeding urban populations, particularly in developing countries. MVAs constitute an "informal" sector, and are rarely subject to official inspection and control. In the end, they can therefore be the source of health safety problems and contribute to the deterioration of environmental hygiene (Botchara., 2013).

1.2 Role of AVP

The informal food sector is a significant source of employment in urban areas, especially for people with low levels of education who might not otherwise find work. In many cities in developing countries ( ), a quarter of the working population, especially women, earn their living from street food sales. Vendors are often women: there is a marked sexual specialization depending on the products sold. Very often, the whole family is involved in the various stages of purchasing products and raw materials, preparing food and selling it (Barron., 2002).

The importance and diversity of MVAs can be explained by the fact that African cities in general, and those of the Democratic Republic of Congo in particular, are becoming melting pots in which people come with their own cultural baggage and eating habits. They look for food suited to their tastes, and in addition to the fast food they offer, vendors play a special role in the local economy (Mughole., 2012).

Many children are also involved in street vending. Many female vendors entered the street food sector primarily to meet their families' food needs (Barron., 2002).

The ease of access and availability of convenience foods is therefore of great interest to consumers. In fact, because of their lower cost, MVAs have become a necessity, thanks to their ability to meet demand for fast food, sometimes without leaving the workplace.

.2. Advantages of AVP

MVAs are an inexpensive, convenient and often nutritious source of food for the urban and rural poor. They are an attractive and varied source for tourists and the well-to-do, a major source of income for large numbers of people, especially women. They offer the possibility of self-employment and an opportunity to acquire business skills for a low to moderate outlay (Lambert., 1989).

.3. Main sales outlets and customers

The range of outlets is extremely diverse, as trade takes place in urban centers.

The classification of sales outlets is based on the criterion of influence and continuous and permanent frequentation of these sites (Mpiana., 2013).

In Kisangani, sales outlets are located in busy areas such as markets, bus stations (parking lots), factories, hospitals, universities and schools. In these cases, vendors usually set up in the same places. Passageways are also excellent selling points, where vendors are more or less mobile (Mughole., 2012).

Spatially, street food flourishes flexibly in areas of high economic activity and demographic concentration. Customers also look for affordable places to eat well, be very comfortable and also be served by vendors, they also look where there is a good dish, place to settle (Rabena., 2011).

In the context of our work, the points of sale were the restaurants (Malewa) in the commune of Kabondo, and the clientele in this environment can be grouped into two categories: a constant clientele made up of sales clerks in the vicinity, agents from the administrative office in this sector, soldiers from Camp Ketele, carpenters, tailors and bicycle and motorcycle repairers; while the other irregular category, no less important, is represented by transporters (motorcyclists, tolekists, etc.), housewives, customers who frequent this area and other sections of the population.), housewives, customers who frequent this sector and other sections of the population.

.4. Constraints related to the preparation and sale of street foods

The lack of official supervision of the street sale of MVAs leads to all kinds of problems directly affecting the health of consumers.

According to Barron (2002), poor food preparation, packaging, preservation and sales techniques in a precarious environment (lack of drinking water, proximity to wastewater drains, sewers and garbage heaps, excessive heat or sunshine) can also be the source of microbiological contamination.

✓ The water

Water is a fundamental element in food preparation. It is indispensable for cooking food and beverages, and for making food edible. It can also be used to improve the nutritional qualities of certain foods (WHO, 1988).

A study carried out by Rozier in 1989 in Pine (India) showed that the drinking water given to their customers by vendors was contaminated with faecal bacteria. This highly contaminated water is probably a major source of diarrhoeal disease. Water pollution can be chemical, physical or faecal (Kazadi., 1998).

One of the key issues for MVAs is the supply of potable water for drinking, washing, cleaning and other operations. The street vendor can only carry a limited supply, and even the fixed vendor does not always have direct access to a supply source.

Indeed, drinking water must meet drinking water standards and be subject to rigorous control. Its bacteriological and hygienic quality depends on its origin, the state of its catchment, the effectiveness of the treatment if it is treated, and the state of the conditions that distribute the water to the tap. Water of dubious hygienic quality is likely to cause microbial contamination of street foods, leading to diseases such as cholera and salmonellosis (Soumare., 1997).

✓ **Waste disposal**

Waste disposal is a common and recent problem in many countries. It takes place in gutters, public dumps or on roadsides. Where there is no drainage system, there is an accumulation of dirty water and garbage, leading to the development of mosquitoes and flies, and serious health problems through the contamination of raw materials and ready-to-eat foods sold in these places (Hobbs, Gilbert., 2001).

In Kabondo, as in other communes in the city of Kisangani, household waste is poorly managed due to the absence of hygiene and sanitation infrastructures, which can lead to a number of harmful phenomena.

Garbage attracts flies, rats, stray cats and dogs, and as it putrefies, it gives off nauseating odours. Some garbage retains water and becomes a breeding ground for mosquitoes.

As a result, they can contain pathogenic germs and parasites that are harmful to living beings. As a result, the accumulation of waste (household refuse) poses not only an environmental pollution problem, but also a public health problem.

The health and nuisance risks associated with waste are mainly due to the proliferation and development of insects and rodents (www.memoireonline.com).

✓ **Personnel and preparation hygiene**

Personal hygiene and food storage hygiene are highly variable, depending on the vendor, with staff and preparation hygiene correlated with water policy. Inadequate hygiene is observed in the cleaning of raw materials and "kitchen" utensils. As a result, if we follow the chain from raw materials to finished products, the problem of strict compliance lies at the root of spoilage and the risk of food poisoning.

Food is often stored at room temperature for long periods. This can lead to the explosive proliferation of germs which cause food to decompose rapidly (www.memoireonline.com).

✓ Prevention

- Avoid micro-organism-induced food spoilage by rapid refrigeration;
- Do not eat food showing signs of spoilage. (www.memoireonline.com)

.5. Microbiology of ready-made meals

✓ **Presentation of ready-made meals**

Prepared dishes play an important role in MVAs. Among the dishes cooked for our work, we distinguish: dishes based on rice, beans and sometimes combined with vegetables (as in the case of pondu).

Depending on how they are presented, ready-to-eat dishes are generally grouped together:

- Hot ready meals;
- Cold ready meals.

Prepared dishes are subject to the action of microorganisms, through food spoilage and deterioration. Sometimes, however, the development of microorganisms on food is not detected either visually or by taste.

As a result, foods can be dangerous. Such is the case with toxins (botulism, staphylococcus, carcinogenic aflatoxins). It should also be noted that certain factors, such as dust and preservation methods, influence the bacteriological quality of food (www.memoireonline.com).

✓ **Food preservation**

Food preservation methods aim to completely eliminate and prevent the development of pathogenic microorganisms. The oldest preservation methods were drying, salting or sugaring, which reduced the water availability of foodstuffs for microorganisms (Inra., 2006).

Delagoutte (2004), nowadays, refrigeration (5°C) and deep-freezing (18°C) are the main methods of food preservation, as they enable longer shelf-life by inhibiting the development of microorganisms.

Cold does not kill microorganisms, but psychrophilic bacteria can grow and spoil food even in cold rooms or refrigerators.

According to Delagoutte, pasteurization and appertization, i.e. the use of high temperatures in food, are designed to kill the vegetative forms of microorganisms, as well as the bacterial forms.

The preservation of food by numerous chemical additives, such as smoking, vinegar preservation and the addition of various salts (nitrite, proportionates, etc.), is proving successful.

✓ **Origin of microbiological contamination**

Potential sources of food contamination include soil, water, plants, animals, people, fertilizers, equipment, ingredients and packaging materials.

Today, street food is a new and fast-growing foodstuff in the urban centers of our countries. Many studies to date have confirmed the risks of contamination and chemical contamination.

These cases of contamination are attributable to poor hygienic practices from the preparation to the packaging, preservation and sale of food in a precarious environment (lack of drinking water, proximity of sewage drains, sewers and garbage heaps, excessive heat or sunshine) may also be the source of microbiological contamination (Kabena., 2011).

.6. The impact of microorganisms in food

The presence of microorganisms in food has the following consequences:

- ✓ Improved food: better preservation and organoleptic quality, thanks to useful flora, manufacturing aids (e.g. yoghurt, cheese, wine, etc.)
- ✓ Food spoilage: deterioration of dietary and organoleptic qualities, due to contamination by common flora: Danger to the consumer: accumulation of pathogenic germs (bacteria, fungi) and their toxins in the body.

.7. Food-borne microbial diseases

Foods can be vectors or actual culture media for microorganisms. They are then potentially capable of causing a variety of ailments in the consumer, the severity of which depends primarily on the nature and number of microorganisms and/or the toxicity of their excretion products.

Among food-borne infectious diseases, the most frequently encountered result from the ingestion of microorganisms belonging to the genera Salmonella, Shigella, Listeria, Brucella, Mycobacterium, Escherichia, Campylobacter, Clostridium, Yersinia, Vibrio and from the ingestion of viruses.

These microorganisms behave like parasites in relation to the body, multiplying by using body components as nutrients. They are invasive and often toxigenic, causing lesions in the digestive tract and other tissues (septicemia).

There are microorganisms that release their toxins into the host organism, such as Clostridium perfringens with toxins A, B, C, D or E produced at stage 3 of sporulation, but also Shigella, *Escherichia coli*, enterotoxinogens with 2 types of toxin (thermostable and thermolabile), *Vibrio parahaemolyticus*, Salmonella endotoxin, etc. These toxins are generally enteric or neurotoxic. These toxins are generally enteric or neurotoxic. (Regnault.,1990)

.8. Main bacterial diseases transmitted

✓ Salmonella infectious disease

The infecting dose with species responsible for serious infectious diseases such as *Salmonella typhi, S. paratyphi A or S. paratyphi* B is just a few cells.

Typhoid fever is often regarded as a disease of dirty hands. Because of the low infectious dose of these Salmonella, the standard for these germs is 0, which justifies all-or-nothing testing, often after a revivification phase (pre-enrichment) followed by a selective enrichment phase. There is, however, the problem of defining the quantity of product to be submitted for analysis in order to meet this standard (Delagoutte, 2004).

✓ Staphylococcal enterotoxicosis

This microbial disease is very common in many countries, particularly France. It results from the consumption of food contaminated with toxigenic strains *of Staphylococcus aureus*. Six types of enterotoxin are currently known (A.B. C. D. E and F); in France, enterotoxin A (65%) is the most frequently encountered, followed by B (20%) and the other types.

Poisoning is characterized by a short incubation period (1 to 4 hours).

Symptoms of this disease, sometimes referred to as banquet disease, include profuse salivation, nausea, vomiting, abdominal pain, profuse diarrhea, sweating, headache, prostration and sometimes fever. Symptoms generally disappear after 24 to 48 hours, and the patient does not develop specific immune defenses (Delagoutte, 2004).

✓ *Clostridium perfringens* toxic infections

This bacterium is probably the most common anaerobic germ found in nature. A saprophyte of soil and water, it is present in many natural products. It is commensal to humans and animals on the skin and in the digestive and even respiratory tracts.

It is thanks to its spore that this bacterium can withstand particularly unfavorable conditions. Its strict anaerobic nature, however, limits its ability to develop in our foods. Canned and cooked foods are excellent culture media for *Clostridium perfringens*, as cooking reduces oxygen levels.

In fact, the spores present on raw meat are resistant to "simmering" for 3 or 4 hours, or to cooking at 110°C for 30 minutes.

A microbial load of at least 108 germs per g is required to trigger toxi-infection.

Symptoms of this disease appear between 8 and 24 hours after consumption of the food. They mainly consist of acute abdominal pain and diarrhea; nausea, vomiting, fever, chills or prostration are rare (Delagoutte, 2004).

- ✓ **Botulinum intoxication**

This poisoning is linked to the ingestion of botulinum toxin, synthesized during the growth of *Clostridium boulinum* in food. This spore-forming, strict anaerobic telluric germ poses a very high contamination risk for many foods. This is particularly true of canned goods (cans and bottles) which are insufficiently heat-treated.

Botulinum toxin" is one of the most violent poisons known: its toxic power is some 500,000 times greater than that of strychnine, and its LD50 (the dose that kills 50% of those who receive it) is estimated at 10-8 to 10-9 g per kg of body weight. For this reason, mortality is high, despite therapeutic measures such as antitoxin serums and anatoxins (Bourlioux., 2000). Given the severity of the intoxication, the hygienic quality of food in this case can only be based on the prevention and control of microbiological quality.

1.9. Other food-related illnesses

Many other microorganisms are implicated in food-borne illnesses.

Although statistically their incidence may be quantitatively insignificant, the fact remains that some of these diseases, sometimes very serious, need to be considered with great care.

Among these, we can mention a few whose germs are responsible: *Shigella*, *Bacillus cereus*, *Listeria*, *Campylobacter*, *Yersinia*, *Vibrio*, *Escherichia coli*, Enteropathogens, Streptococcus A and D... (Bourlioux, 2000).

There are many food-borne diseases that have a detrimental effect on human health. These diseases are transmitted by vectors (flies, insects, etc.) which circulate freely from a contaminated area to an uncontaminated one: but also by lack of hygiene.

✓ **Food poisoning**

Toxi-infections are the most common form of food poisoning.

The main germs responsible for poisoning are grouped into two categories according to criteria:

- According to the criterion of adhesion and penetration of the intestinal mucosa, we count bacteria belonging mainly to the germs Salmonella, Shigella, *Escherichia coli*, Vibrio parahemolyticus, Campylobacter, Listeria.

- Those that act via a toxin of the Staphylococcus, Clostridium or Bacillus genus (www.memoireonline.com).

1.10. Genus Escherichia

Escherichia coli causes urinary tract infections, wound infections, appendicitis, peritonitis, choleocystitis, bacteremia and meningitis, especially in newborns.

Four groups of Escherichia coli are responsible for diarrhea, particularly in children, but also in adults.

- ETEC: (enterotoxigenic E. coli): these strains cause severe enteritis complicated by dehydration, often fatal in infants and young children;
- EIEC (enteroinvasive E. coli): these immobile strains invade the ileal and colonic mucosa, causing dysentery with pus and mucus, as in shigellosis. Their biochemical reaction resembles that of shigella.
- EPEC (enteropathogenic E. coli): these strains are associated with diarrhea in infants and sometimes in adults.
- EHEC (enterohemorrhagic E.coli): severe, profuse bleeding (www.memoireonline.com).

1.11. Total aerobic mesophilic flora (FMAT)

According to http: en.wikipedia.org/wiki/organisme, total aerobic mesophilic flora (TAMF) is a sanitary indicator that enables us to evaluate the number of CFU (colony-forming units) present in a product or on a surface. This count is carried out at 30°C, which enables three main types of flora to be counted:

- Thermophilic flora, optimum growth temperature 45°C;

- Mesophilic flora, optimum growth temperature between 20 and 40°C;

- Psychrophilic flora, optimum growth temperature 20°C.

A colony-forming unit (CFU) is an observable colony on agar, which may come from an isolated microorganism, a spore or a combination of microorganisms.

However, the multiplication of FMAT in street foods can indicate the state of freshness and general hygiene of the food. On markets, food products can be contaminated by staphylococci, due to the action of factors such as: wind, dust and contamination of human origin through the handling of materials, salivary secretions and sometimes even sweat secretion (Daouda, 2002).

1.12. The staphylococcus genus

Staphylococci are defined as Gram-positive, immobile, non-spore-forming, catalase-positive and oxidase-negative cocci.

These bacteria are ubiquitous: found in air, soil and water, they are part of the commensal flora of the skin and mucous membranes of humans and animals (Cuq, 2007). They belong to the Micrococcaceae family, and vary in diameter from 0.5 to 1.5 m (Monica, 2000).

Three species have been identified:

✓ *Staphylococcus*

Causes a wide variety of necrotic diseases such as furunculosis, anthrax, abscesses, panicitis, phlegmon and wound infections; secretes coagulase and forms a golden pigment, ferments mannitol;

- ✓ *Staphylococcus*

A constant commensal of the skin, it is usually pigmented white or yellow, occasionally orange, does not produce coagulase and does not use mannitol as a fermentable substance for anaerobic growth (Lambert, 1989).

- ✓ *Staphylococcus*

Develops on dead tissue, it does not produce coagulasse, does not ferment mannitol and is generally white-pigmented.

These last two species were long considered non-pathogenic, but can nevertheless cause suppurative infections and even septicemia when conditions are favorable for their proliferation (Margairaz et al., 1975).

1.13. Enterobacteriaceae

Enterobacteriaceae are Gram-negative bacteria, mobile or immobile, aerobic and aerobicosis.

They belong to the phylum Eubacteria, class Asporutales, order Bacteriales, family Enterobacteriaceae (Monica., 2000).

There are 28 genera, including 14 of medical interest (Citrobacter, Enterobacter, E. coli, Klebsiella, Morganella, Proteus, Providencia, Salmonella, Shigella, Yersinia, Serrasia, etc.).

They are commonly found in the intestines of humans and animals, or in uncontaminated soil and air.

A large number of Gram-negative bacteria have been reported to cause foodborne gastroenteritis. However, the most important of these is Salmonella. In addition to Salmonella, we also have Shigella, coliforms and halophilic Gram-negative rods (Lambert., 1989).

Symptoms caused by these germs include vomiting, diarrhea, nausea and abdominal pain. Preventive measures must be taken at various levels: slaughtering, transport to market, handling, sale, etc. (Lambert, 1989; Gentilini, 1993).

1.14. **Salmonella**

Salmonella are enterobacteria with the essential characteristics of not fermenting lactose and not producing urease; they are parasites of humans, mammals (rodents), birds and cold-blooded animals (reptiles); after oral penetration, they are responsible for numerous infections (Salmonellosis), including typhoid and paratyphoid fevers, gastroenteritis and collective food poisoning.

The main mode of human contamination is ingestion from water (especially *S. typhi*), food (e.g. dairy products, eggs, meat) or carrier pets (Etobo., 2006).

The Salmonella genus comprises just one species, *Salmonella enterica*, which includes subspecies subdivided into around 2,000 on the basis of their 0, H and capsule antigens (Etobo., 2006).

Here are just a few of the salmonella responsible for human disease:

- *S. typhi*: causes typhoid fever;
- *S. paratyphi*: responsible for paratyphoid fever;
- S. *tphimerium*: responsible for food poisoning;
- S. cholerae-suis: responsible for food poisoning;
- *S. gallinarium*: responsible for avian dysentery.

1.15. Origin of microbial contamination of foodstuffs

Microbial contamination of food can originate from a variety of sources. These microscopic contaminants include bacteria, viruses, fungi and parasites. Their presence in food can lead to foodborne illness, also known as food poisoning.

The main sources of microbial contamination of foodstuffs are as follows:

✓ Raw materials

Food can be contaminated as soon as it is produced, whether during harvesting (e.g. through contact with animal excrement) or during the rearing of animals for consumption (e.g. in the case of contamination of meat with bacteria such as salmonella).

✓ The water

Water used for food washing or as an ingredient in food preparation can contain pathogenic micro-organisms. If water is contaminated with bacteria such as Escherichia coli or parasites such as Giardia, these contaminants can spread to food.

✓ Handling and production environment

Poor hygiene practices when handling food, such as failure to observe hand-washing rules, can lead to microbial contamination. In addition, production facilities and equipment must be properly maintained and cleaned to avoid cross-contamination between foods.

✓ Storage and preservation

Poor food preservation, such as keeping temperatures too high, can encourage the growth of pathogenic micro-organisms. Bacteria such as Salmonella and Listeria can multiply rapidly when exposed to temperatures between 4°C and 60°C.

✓ Transportation

Food transport can also be a source of contamination. If transport conditions are inadequate, for example by maintaining inappropriate temperatures, micro-organisms can grow and contaminate food.

Table 1: Origin of food poisoning www.memoireonline.com)

Type	Origin	Prevention
Shigella, E. Colientero-Invasive	Undercooked or uncooked food, contaminated during preparation	Sufficient cooking
Compylobacter jejuni	Main source: poultry milk and water, germ present in the digestive tract of most domestic animals	Storage at 4°C sufficiently cooked
Yersimiaentero litica	Full-fat dairy products, ice cream, seafood, butcher's meat and offal (pork tongue, pre-cooked or pre-cut vegetables)	Preservation when sufficiently cooked
Salmonella sp	Meat, poultry, eggs, pastries, ice cream, fish	Preservation when sufficiently cooked
Staphylococcus aureus Bacillus cereus	Contaminated ready-made milk and dairy products not kept refrigerated	Preservation when sufficiently cooked
CI. Botulinum Cl. Perfringens	Pork and cold cuts, Poorly sterilized family preserves	Sterilization efficiency Family preservation

CHAPTER TWO: MATERIALS AND METHODS

2.1. Study environment

This study was carried out in the city of Kisangani, capital of Tshopo province (Figure 1). It is a city, which is located in the central Congolese basin, at 0° 31′ 00″ North and 25° 11′ 00″ East. Its average altitude is 396 m (Nyakabwa, 1982). Administratively, Kisangani comprises 6 communes: Kisangani, Makiso, Mangobo, Kabondo, Tshopo and Lubunga, covering a total area of 1,910 km^2 (Etobo, 2012).

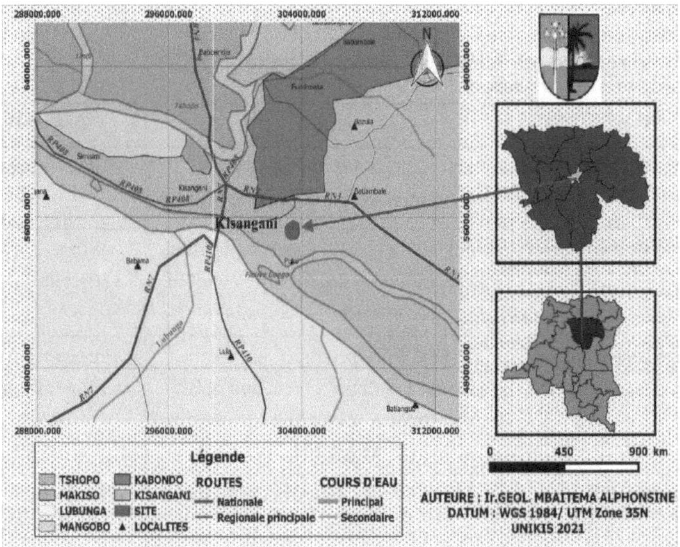

Figure 1: Map of the city of Kisangani.

2.2 MATERIALS

2.2.1. Biological materials

Our study material consisted of foodstuffs sold along the Camp Ketele road (Malewa: Rice-Beans-Pondu) in the commune of Kabondo in Kisangani: more precisely, in the three restaurants (Malewa) located on either side of the main Camp Ketele road; then brought back to the microbiology and phytopathology laboratory of the Faculty of Science at the University of Kisangani (Unikis).

2. 3. Sample collection

Samples were collected at three different sites during the period August-October 2023. Samples were taken at random, but within the same area. The harvested samples were transported in sterilized jars, then taken back to the Faculty of Science's microbiology and phytopathology laboratory for analysis.

2.4. Methods

Microbiological analysis of samples is carried out in four stages: preparation of dilutions, inoculation into culture media, enumeration, characterization and identification of microorganisms.

2.5. Microbiological analysis

In a jar containing 225 ml of physiological water, add 25 g of food sample, mix the solution, and the resulting solution is called the stock solution.

It corresponds to dilution 10^{-1}.

This step enables the micro-organisms to be revitalized. Starting with dilution 10^{-1}, successive dilutions were made in test tubes each containing 9ml of physiological water. Using a sterile syringe, we then took 1ml from the first dilution and added it to the second tube. This gave us dilution 10^{-2}. From this, we repeat the same operations to successively obtain dilutions 10^{-3}, $10^{(-4)}$ and 10^{-5}.

The dilutions are then inoculated onto different culture media for the enumeration of undesirable germs. This will enable us to assess the level of contamination of our samples by undesirable germs such as: total mesophilic flora, total coliforms, *Staphylococcus aureus*, Salmonella, etc.

2.5.1. Determination of bacterial load, FMAT at 30°C

Take 1ml of dilution $10^{(-4)}$), 10^{-5} and pour aseptically into each empty, sterile petri dish.

Next, pour 15 ml of supercooled nutrient agar at 50°C into the same dish. Homogenize the solution and wait for the medium to solidify completely.

Invert the dish before placing in the incubator at 37°C for 72 hours. Counting is done by counting colonies. Viable plates are those with colonies between 30 and 300.

2.5.2. Enumeration of *Staphylococcus sp* at 37°C

1 ml of the $10^{(-1)}$) stock dilution is introduced into each Petri dish. Same procedure as above (2.2.2.1). Staphylococci are counted on Chapman's medium (MSA, Mannitol Salt Agar) containing a high concentration of NaCI (75%) tolerated only by staphylococci. Turn the dish upside down and incubate at 37°C for 24 to 48 hours. Counting is by colony count for plates between 15 and 300 colonies.

2.5.3. Coliform enumeration

1ml is taken from the $10^{(-1)}$) stock solution and introduced into each Petri dish, to which EMB agar is added. Homogenize and allow the medium to solidify.

After solidification, turn the plates upside down and incubate at 44°C for fecal coliforms and 37°C for total coliforms (in our case). Incubation lasts 24 to 48 hours.

Counting is by colony count for plates between 15 and 300 colonies. Only black or grey colonies are counted.

2.5.4. Salmonella enumeration

Take 1ml of the 10^{-1} ml dilution and enrich in a test tube containing Selenite broth. Incubate at 37°C for 24 hours. Then take 1ml of the selenite broth culture and place in an empty sterile Petri dish. Aseptically pour supercooled SS Agar medium into the same dish. Homogenize and allow medium to solidify. Incubation took place in an oven at 37°C for 24 to 48 hours. Plates are placed upside down (lids down) to avoid confluence of surface colonies due to condensation of water on the lid. Counting is done by colony counting for plates between 30 and 300 colonies.

Colony count results are expressed in Colony Forming Units per gram (CFU/g). To obtain the exact number of germs, the number of colonies counted is multiplied by the dilution factor, which corresponds to the inverse of the dilution considered.

2.6. Strain preservation

For preservation, the various strains were deep-pitted in soft agar.

2.7. Strain characteristics

2.7.1. Staphylococcus

2.7.1.1. Morphological characteristics

✓ **Differential Gram staining**

This is a complex staining procedure that is mandatory for the initial identification of microorganisms.

The various stages of Gram staining can be summarized as follows

- A pure, isolated bacterial colony is removed from the Petri dish using a platinum loop.
- A thin smear is made on a slide with the loop. The slide is then rapidly fixed above the Bunsen burner flame.

For

- A few drops of gentian violet are poured onto the slide and left to act for a few minutes.

  60 seconds. The blade is then washed under running water.
- A few drops of lugol are poured onto the slide to react for 30 seconds. The slide is again washed under running water.
- The slide is coated with ethanol (alcohol) for 10 seconds, then washed immediately to avoid excessive discoloration.
- Fuchsin is applied for 10 seconds to recolour the sample.
- The blade is dried on the Bunsen burner flame.

- The slide is coated with immersion oil before observation under an optical microscope with X100 objective (Mombeni, 2022).

2.7.1.2. Biochemical characteristics

The strains to be studied were seeded in physiological water and incubated in an oven at 37°C for 24 hours.

This culture is used to perform various tests.

✓ **Catalase test**

Detecting the presence of catalase in bacteria is essential to differentiate *Staphylococcus* (catalase positive) from *Streptococcus* (catalase negative).

Although it is primarily useful for differentiating between these genera, it is also useful in distinguishing between species belonging to the same genus such as *Aerococcus* (catalase positive) and *Aerococcus viridians* (catalase negative).

The catalase test is also useful for differentiating between obligate aerobic and anaerobic bacteria.

- Place a drop of hydrogen peroxide on a clean, dry slide,
- Using a buttoned Pasteur pipette, add bacterial inoculum,
- Observe immediately,

  If bubbles appear, release of dioxygen: **catalase**

- If bubbles are absent: **negative catalase** (Prudencio,2018).

✓ **Coagulase test**

The test for the ability of Staphylococci to coagulate plasma is the main test characterizing *S. aureus*. This detection test consists of incubating for at least 4 h at 37

°C a 0.5 ml mixture of plasma and test strain. The appearance of a clot is observed after 4 hours by tilting at 90°.

If no clot is observed after 4 hours, the assay will be continued with overnight incubation at room temperature and final observation at 24 hours (Mekhloufi., 2018).

This test identifies 99% of S aureus strains, but some strains do not produce coagulase. In such cases, other tests are used to identify the species.

2.8.Biochemical and physiological characteristics

Isolated strains were transferred to tubes each containing 3 ml of peptone water (10g/1L) for revivification, then incubated in an oven at 37° C for 24 hours to perform various biochemical tests.

2.8.1. Kligler midfield

It demonstrates the rapid formation of lactose and glucose with or without the production of gas and hydrogen sulfide. Strains are picked up using a sterile platinum loop and inoculated onto a sloping slope of Kliger medium in a test tube using the serial streak method, then incubated at 37°C for 24 hours.

Kligler's medium gives us 4 different results:

- ✓ Glucose fermentation:

- Red cap: unfermented glucose (-)
- Yellow cap: fermented glucose (+).

- ✓ Lactose fermentation

Red slope: unfermented lactose (-)

Yellow slope: fermented lactose (+)

- ✓ Gas production:

Appearance of voids in the tube

- ✓ Formation of hydrogen sulfide:

Formation of a black color along the sting

Procedure

- o From a suspect colony collected on a selective isolation medium, perform a 24-hour preculture, then inoculate the suspension with a central prick and tight ridges on the sloping surface
- o Incubate at 37°C for 24 hours, capsules loosened, to promote gas exchange.

✓ Reading

	Interpretation of results
Glucose	If the pellet turns yellow: Glucose is positive (+), otherwise negative (-).
Lactose	If the slope turns yellow: lactose is positive (+), otherwise it is negative (-).
Gas	Appearance of gas bubbles in the pellet (gas +), otherwise the result is negative
H_2S	Formation of a black ring between base and slope (H_2S +), otherwise (H_2S -)

Table 2: Kligler test reading

2.8.2. MIU medium (indole urease mobility)

This environment highlights the following characteristics

- ✓ Presence of urease
- ✓ Presence of tryptophanase
- ✓ Presence of tryptophan deaminase (TDA)
- ✓ Mobility or motility.

A. Principle

Bacteria possessing urease convert urea into ammonium carbonate, resulting in alkalinization and a purplish-red coloration of the medium in the presence of phenol red (pH indicator). Indole production is highlighted by the addition of Kovacs reagent, which interacts with indole to produce a red coloration in the upper part of the medium in the event of a positive reaction. The presence of tryptophan deaminase (TDA) is demonstrated by the addition of iron perchloride, which produces a red-brown coloration of the medium in the event of a positive reaction.

B. Procedure

- ✓ From a suspect colony collected on a selective isolation medium, perform a 24h preculture, then inoculate a few drops of the suspension by deep pricking;
- ✓ Incubate tubes at 37°C for 24 hours.

C. Reading

Table 3: MIU test reading

	Interpretation of results
Urease	If medium turns purplish red: urease (+), otherwise urease (-)
Indole	If there is a red coloration on the surface of the medium: indole (+), otherwise (-).
TDA	If the middle color turns reddish brown: TDA (+), if orangey yellow: TDA (-)
Mobility	If the microbial mass invades the entire tube, mobile (+), if it develops along the line of the puncture (±), if it develops only along the line of the puncture, immobile (-).

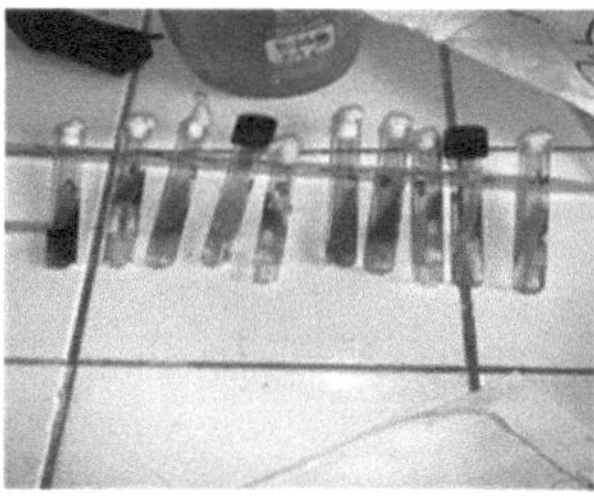

Fig.3: Kligler test

Fig.4: MIU test

2.8.4. Simmons Citrate medium

This is an agar medium used for the identification of Enterobacteriaceae using citrate as the sole carbon source.

✓ Principle

The use of citrate results in alkalinization (+), with the medium turning from green to brilliant blue. Otherwise, there is no color change. Escherichia coli (including serotypes responsible for infantile enteritis), *Shigella, Yersinia* and *Edwardsiella* do not grow on this medium. *Serratia*, most *Enterobacter, Citrobacter, Klebsiella, Proteus* and *Providentia*, (except

Morganella morganii and *Klebsiella rhinoscleromatis*) use citrate and give a characteristic blue hue. Simmons citrate agar can be used to differentiate *Salmonella Enteritidis*, which are citrate (+), from *Salmonella Typhi*, *Paratyphi* A, Pullorum and Gallinarum, which are citrate (-).

- ✓ How it works
  - o Prepare inclined agar tubes and inoculate lightly on the surface of the medium;
  - o Incubate at 37°C for 48 hours.
- ✓ Reading

Table 5: Simmons citrate test reading

	Interpretation of results
Citrate	- If the medium turns blue: citrate (+) - If the medium does not turn, remains green: citrate (-)

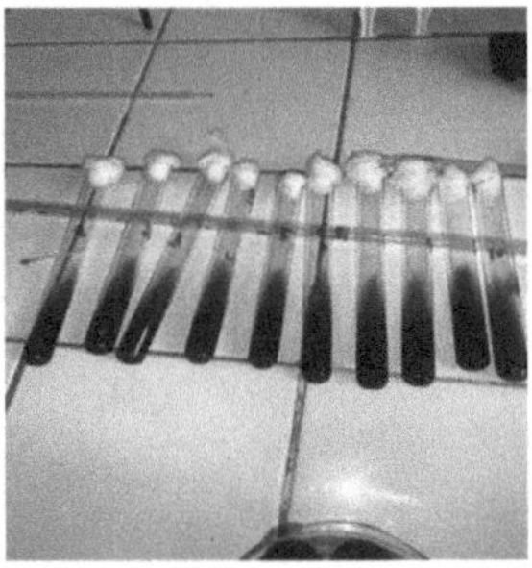

Fig.4. Simmons citrate test

2.9. Studying the sensitivity of germs to antibiotics

This method involves aseptically depositing solution-soaked disks onto the agar using sterile forceps. The discs must adhere well to the agar and be placed in such a way as to avoid overlapping the zone of inhibition, so that the distance between the center of the disc is 30mm and that between the edge of the petri dish and the center of the disc is 15mm.

The dishes are then left for 15 minutes at laboratory temperature to allow the antibiotics to diffuse and are turned upside down before being incubated in an oven for 10 to 24 hours at 37° C (BIOMERIEUX, 1980, Institut Pasteur. 1983).

The antibiotic contained in the disc will diffuse along a gradient of concentrations greater than or equal to the minimum inhibitory concentration.

The result is a zone of inhibition around a disc that varies in size depending on the strain's sensitivity and the antibiotic's ability to spread.

2.9.1. Reading the sensitivity test

The effect of antibiotics on germs is demonstrated by the appearance of zones of inhibition. A zone of inhibition is defined as The clear halo around discs where there is a total absence of growth.

Bacterial sensitivity to antibiotics is assessed by measuring the zone diameter with a lath. The value obtained is comparable to the antibiotic's critical diameter:

- Diameter of inhibition zone less than critical diameter: conclude resistant (R)
- Diameter of inhibition zone greater than critical diameters: conclude sensitive (S)
- Intermediate responses are considered resistant (Institut Pasteur. 1983).

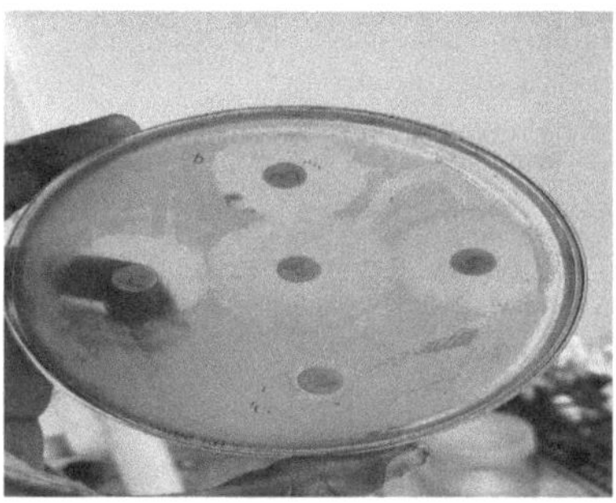

Fig.5 Bacterial activity with the formation of halos around the discs

2.9.2. Critical inhibition zone diameters (in mm)

Table 4 below shows the critical diameters of some antibiotics.

N°	Antibiotics	Resistant	Intermediate	Sensitive
1	Amoxicillin	≤ 13	14 - 18	≥ 19
1	Ciprofloxacin	≤ 13	14 - 20	≥ 21
2	Doxycycline	≤ 14	17 - 18	≥ 19
4	Erythromycin	≤ 16	17 - 21	≥ 22
5	Gentamycin	≤ 16	17 - 18	≥ 19

Table 6: Critical diameters of some antibiotics in millimeters (John *et al.*, 197

2.9.3. Expression of results

Colony count results are expressed in colony-format units per gram (CFU/g). The number of colonies per gram was calculated using the formula below (Martial, 2016).

$$N = \frac{£c}{(n1+n2)\ d.v}$$

With

£c: sum of colonies

d: The dilution rate of the first dilution used for the box counts

n1: Number of boxes of the lowest dilution

n2: Number of boxes of the highest dilution

V: The volume of inoculum applied to each box (usually 1ml by mass and 0.1ml by surface)

N: Number of colonies in CFU/g

2.9.4 Interpretation of results

The microbiological quality of samples is assessed according to a three-class plan (AFNOR, 1996). For this classification, m is considered the microbiological criterion. Thus, a product will be classified as:

- **Satisfactory** when results are ≤ 3m
- **Acceptable** when results are > 3m and ≤ 10m
- **Unsatisfactory** when results > 10m

These m criteria are expressed in CFU/g.

2.10. Statistical analysis of analysis data

In order to compare the various target sites in relation to the mean values of the microbiological parameters considered, a descriptive analysis was carried out. To compare the variation considered from one site to another, an analysis of variance (ANOVA) was performed.

2.11. Deontology and ethics

The data was collected anonymously with the informed consent of the sellers.

CHAPTER THREE: RESULTS AND DISCUSSION S

In this chapter, we present and discuss the various results obtained from the microbiological analysis of samples of "Rice, beans and pondu" sold at three sites (Malewa) in the commune of Kabondo, located close to each other: Malewa 1(M1), Malewa 2(M2) and Malewa 3(M3).

Bacterial loads are presented first, then the results are compared with the standards required for the hygienic quality of food sold on public highways. This is followed by a comparative analysis of three sites considered as a whole, and finally we present the results of tests to characterize the germs sought.

3.1. BACTERIAL LOAD OF SAMPLES

3.1.1. Samples from the 1st restaurant (Malewa 1)

The bacterial loads of samples from the first restaurant are highlighted in Table 6.

Table 6. Bacterial loads of rice, bean and pumice samples from the first restaurant (Malewa 1). Values are expressed in CFU/g of food. CT: total coliforms, S. aureus: *Staphylococcus aureus*, Ech:

Sample	FMAT	C.T	*S. aureus*	*Salmonella*
Ech 1	$80.\ 10^5$	$4.\ 10^2$	$33.\ 10^2$	4
Ech 2	$23.\ 10^5$	1	$2.\ 10^2$	1
Ech 3	$70.\ 10^5$	2.10^2	$17.\ 10^2$	2
Average	$57,6.10^5$	$2,3.10^2$	$25.\ 10^2$	2,3

The average loads obtained for FMAT, fecal coliforms, Staphylococci and Salmonella were $57.6.10^2$, $2.3.10^2$, 17.3. 10^2 CFU/g and 2.3 CFU 10^2/g respectively.

The highest load was observed for FMAT ($57.6.10^5$ CFU/g) and the lowest for Salmonella (2.3 CFU/25g). The majority of samples showed the presence of the desired germs in abundance, with the exception of sample E2, where total coliforms and Salmonella were found in abundance (1 colony observed).

3.1.2. Samples from the 2nd restaurant (Malewa: M2)

The bacterial loads of samples from the second restaurant are highlighted in Table 7.

Table 7. Bacterial loads of rice, bean and pumice samples from the second restaurant (Malewa 2). Values are expressed in CFU/g of food.

Sample	FMAT	C.T	S.aureus	Salmonella
Ech 1	$20.\ 10^5$	$6.\ 10^2$	$35.\ 10^2$	2
Ech 2	$70.\ 10^5$	$3.\ 10^2$	$23.\ 10^2$	6
Ech 3	130.10^5	$2.\ 10^2$	$30.\ 10^2$	3
Average	$73,3.10^5$	$3,6.10^2$	$29.\ 10^2$	3,6

The average loads obtained from samples from the second restaurant appeared higher than those from the first. The average loads obtained for FMAT, fecal coliforms, Staphylococci and Salmonella were $73.3.10^5$, $3.6.10^2$, 29. 10^2 CFU /g and 3.6 CFU/25g of food respectively. The highest load was observed for FMAT ($73.3.10^5$ CFU /g) and the lowest for Coliforms and Salmonella, which were equal (3.6 CFU/25g). The majority of samples showed the presence of the germs we were looking for in abundance, with the exception of sample E2, where total coliforms and Salmonella were found in small quantities (1 colony observed).

3.1.3. Samples from the 3rd restaurant (Malewa: M3)

The bacterial loads of samples from the third restaurant are highlighted in Table 8.

Table 8. Bacterial loads of rice, bean and pumice samples from the third restaurant (Malewa 3). Values are expressed in CFU/g of food.

Sample	FMAT	C.T	S.aureus	Salmonella
Ech 1	$17.\ 10^5$	$3.\ 10^2$	$25.\ 10^2$	2
Ech 2	50.10^5	0	$3.\ 10^2$	0
Ech 3	$80.\ 10^5$	2.10^2	$15.\ 10^2$	3
Average	$49.\ 10^5$	$1,6.\ 10^2$	$14,3.\ 10^2$	1,6

The average loads obtained from samples from the third restaurant appeared to be lower than those from the first and second restaurants. The average loads obtained for FMAT, faecal coliforms, Staphylococci and Salmonella were 49.10^5, $1.6.10^2$, 14.3. 10^2 CFU/g and 1.6 CFU/25g of food respectively. The highest load was observed for FMAT (49.10^5 CFU /g) and the lowest for Salmonella (1.6 CFU/25g). The majority of samples showed the presence of the desired germs in abundance, with the exception of sample E2, where total coliforms and Salmonella were absent (i.e. 0 colonies observed).

All samples from the first and second restaurants showed germs in rice, beans and pondu (Tables 6 and 7). This was in contrast to samples from the third restaurant, where there were some exceptions (Table 8).

3.2. HYGIENIC QUALITY OF FOOD SAMPLES

To assess the quality of the rice, bean and poppy samples, the bacterial loads obtained were compared with the microbiological criteria. The international standards of the Agence Française de Normalisation (AFNOR, 1996) were applied. Samples were judged satisfactory, acceptable and unsatisfactory according to the three classes defined by the criteria (Cfr Chapter II. Materials and methods, point 2.7). The hygienic quality of rice, bean and egg samples from three restaurants (Malewa: M1, M2 and M3) is given in Tables 9 and 10.

3.2.1. Hygienic quality of samples from the 1st and 3rd restaurants (Malewa: M1 and M3) in the commune of Kabondo

The hygienic quality of samples from the first and third restaurants is shown in Table 9.

Table 9. Hygienic quality of rice, bean and spawn samples from the first and third restaurants (Malewa: M1 and M3). FMAT, Total aerobic mesophilic flora; TC, Total coliforms; No. of Ech, Number of samples analyzed.

Germs	Satisfaction		Acceptale		No satisfaction	
	No. d'Ech	%	No. of Ech	%	No. of Ech	%
FMAT	0	0	0	0	3	100
CT	1	25,1	0	0	2	74,9
S. aureus	0	0	0	0	3	100
Salmonella	1	25,1	0	0	2	74,9

All samples from the first and third restaurants, i.e. 100%, were classified as unsatisfactory for FMAT and Staphylococcus (*S. aureus*). 74.9% were unsatisfactory for fecal coliforms and *Salmonella.* On the other hand, 25.1% of samples met the criteria. However, no samples were declared acceptable.

3.2.2. Hygienic quality of samples from the 2th

In contrast to the first restaurant, all samples taken together from the second restaurant failed to meet quality standards. No sample was found to be satisfactory, let alone acceptable. This situation is illustrated in Table 9 below.

Table 9. Hygienic quality of rice, bean and pondu samples taken at the second restaurant. FMAT, Total aerobic mesophilic flora; CF, Faecal coliforms 44°C (E. coli); No. of Ech, Number of samples analyzed.

Germs	**Satisfaction**		**Acceptale**		**No satisfaction**	
	No. d'Ech	%	No. of Ech	%	No. of Ech	%
FMAT	0	0	0	0	3	100
CT	0	0	0	0	3	100
S.aureus	0	0	0	0	3	100
Salmonella	0	0	0	0	3	100

The results show that the majority of samples analyzed are contaminated with total aerobic mesophilic flora (TAMF), fecal coliforms (FC), Staphylococci and Salmonella, with loads exceeding the limits set by standards.

FMAT is an alteration flora. It is made up of non-specific germs, which are considered to be hygiene test germs. High FMAT loads indicate ineffective processing or preservation for food samples sold in the vicinity of markets in the city of Kisangani (Ezechiel et *al.*, 2012).

Total and faecal coliforms are indicators of faecal and food contamination. Here, too, the loads obtained far exceed the required standards. Their presence in large numbers leads to the suspicion of Salmonella in the samples analyzed. We believe that these high coliform loads in the samples analyzed can be explained by poor hygiene practices (non-washing and non-disinfection of hands or after handling contaminated packaging, etc.) among sellers of caramelized peanuts (Euloge et *al*., 2012).

Swedi (2021) in a study carried out among sellers of roasted and caramelized peanuts in the city of Kisangani found very high FMAT values of up to $85.6.10^6$CFU/g and 3.10^2 CFU/g for coliforms. They attributed this contamination to the storage process, i.e. the conditions and duration of peanut storage, handling and transport.

The food samples analyzed contained staphylococci and Salmonella. Pathogenic staphylococci are represented by *Staphylococcus aureus*, a germ of human contamination. The coagulase test performed was positive for all ten strains isolated on Mannitol Salt Agar (MSA). *S. aureus* can be used to determine which products present the greatest risk of intoxication. *S. aureus* and *Salmonella* are potentially pathogenic germs (Ambegebo, 2016). In markets, food products can be contaminated by staphylococci, due to the action of factors such as: wind, dust and contamination of human origin through the handling of materials, salivary secretions and sometimes even sweat secretion (Daouda, 2002).

Among the Salmonella species, the most dangerous is *Salmonella typhii*, which represents a real danger to consumer health. Their presence in our samples testifies to the unhealthiness of the rice, beans and pudding sold in the public streets of the city of Kisangani. It should be remembered that according to quality criteria (AFNOR, 1996), the number of pathogenic colonies tolerated in a food product must be zero in 25 g for *Salmonella*. The presence of these germs in foodstuffs is sufficient to classify them as unsatisfactory (Diouf, 1992; WHO, 2003).

On the whole, the samples analyzed were of unsatisfactory hygienic quality, contaminated with high levels of bacterial germs that far exceeded the thresholds required by quality standards. The detection of potentially pathogenic germs (*S. aureus* and *Salmonella*) in these samples reveals that the food sold on public roads is of poor quality and constitutes a health risk for the population of the city of Kisangani.

We believe that this contamination stems from a number of unhygienic conditions: handling during preparation, and especially when serving food to customers, as vendors do not observe hygiene rules when serving. Poor storage conditions and unsanitary conditions in the vicinity of the restaurants (M1, M2 and M3) also contribute to this contamination.

3.3. COMPARISON OF SAMPLES AT THREE RESTAURANTS

The average charges obtained differ between the three restaurants. In fact, they were higher in the second restaurant (M2) than in the other two (M1 and M3). 100% of samples from the first and second restaurants revealed the germs we were looking for. This was not the case for samples from the third restaurant, where a few exceptions were noted. These were the absence of faecal coliforms and *Salmonella* in Sample 2 from the third restaurant (i.e. 0 colonies observed) (Table 7). In terms of hygienic quality, no sample from the second restaurant was found to be either satisfactory or acceptable.

Figure 6 shows the variation in average bacterial loads observed at the three sites.

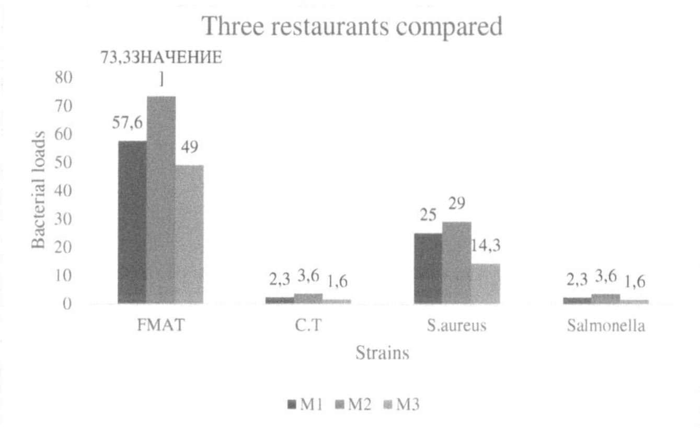

Figure 6. Variation in average germ loads of rice, bean and poppy feed samples from Kabondo restaurants

3.4. Comparative analysis of three market samples

The ANOVA test reveals a significant difference between the samples from three restaurants in the commune of Kabondo ($p<0.05$).

3.5. Characterization and identification of isolated strains

Results on the characterization and identification of isolated strains are illustrated in figure 7 below:

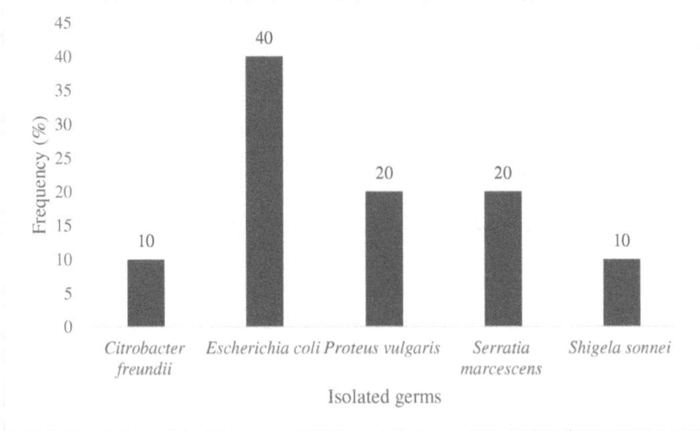

Figure 7. Load variation of isolated strains

Figure 7 shows, with the help of identification key, that there are 5 different strains of bacteria identified in the samples analyzed. *Escherichia coli* species are the most frequent, with a frequency of 40%, followed by *Proteus vulgaris* and *Serratia marcescens* with 20% each. On the other hand, *Citrobacter freundii* and *Shigela sonnei* are the least frequent strains, with a frequency of 10% each.

Our results differ from those of Abedi (2021), who found 5 different strains of bacteria, the most frequent of which had a frequency of and the least frequent . Our results also differ from those of Nyamanlifofe (2016), who found that of strains isolated from all three sources were *Klebsiella* sp and only were *Enterobacter* sp.

3.6. Strain sensitivity test

Results for antibiotic susceptibility of isolated strains are shown in figure 8.

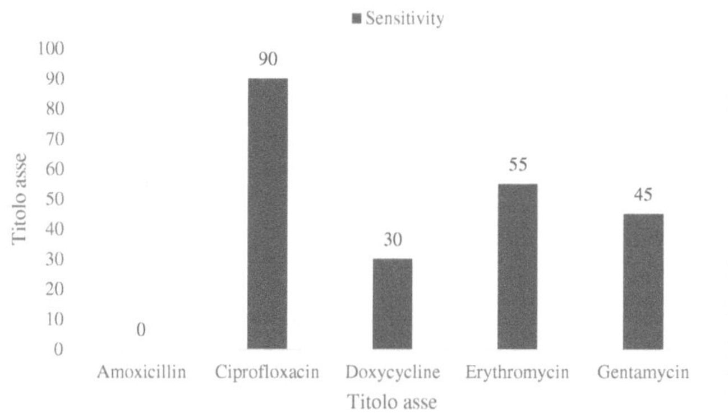

Figure 8. Antibiotic susceptibility of isolated strains

Figure 8 shows that of strains tested are sensitive to Ciprofloxacin, 55% of strains tested are sensitive to Erythromycin, 45% of strains tested are sensitive to Gentamycin and only 30% of strains tested are sensitive to Gentamycin. And Amoxicillin has 0% sensitive antibacterial activity to all strains tested. Furthermore, all strains tested are 100% resistant to Amoxicillin.

By comparing our results with those of Nyamaifofe (2015), who found 50% of them sensitive to Ciprofloxacin, our results are similar. Our results are also similar to those of Abisa (2004), who found that 50% were sensitive to the same antibiotic.

Our results on Erythromycin are almost identical to those of Abedi (2021), who found 70% of the strains tested to be resistant to Erythromycin.

CONCLUSION AND SUGGESTIONS

This work is a contribution to the study of the microbiological quality of food sold on public roads (AVP). Its aim was to carry out a bacteriological analysis of food sold at three sites (Malewa) in the commune of Kabondo, namely Malewa 1 (M1), Malewa 2 (M2) and Malewa 3 (M3). Thus named in this work.

The hypotheses were that the hygienic quality of street food in Kisangani would be unfit for consumption, and that the bacterial load of food hygiene indicators (FMAT), Staphylococci and Total Coliforms would differ according to the food group considered. And would be responsible for food poisoning in consumers, these strains will be resistant to antibiotics commonly sold in Kisangani and the consuming population will be exposed to a risk.

Samples taken from these three sites were subjected to microbiological analysis for Total Aerobic Mesophilic Flora (TAMF), total coliforms (in our case), *Staphylococcus aureus* and Salmonella. Germ characterization using biochemical, physiological and morphological tests enabled presumptive strain identification.

The results show the presence of spoilage germs (FMAT), contamination indicators (fecal coliforms and total coliforms in our case) and potentially pathogenic germs (Salmonella, Staphylococcus aureus) in the majority of samples from three sites.

The evaluation of the hygienic quality of the samples revealed the following:

- FMAT load: of samples were satisfactory for all three sites,

33% acceptable and 11.unsatisfactory,

- faecal coliform load: 55.satisfactory and 44.acceptable,

- Staphylococcus aureus load: 100% of samples acceptable for all three sites,

- Salmonella load: only 11.2% satisfactory and 88.unsatisfactory, representing a potential health risk for consumers in particular and the urban population in general.

Samples from the second restaurant in the commune of Kabondo are of poor quality compared with those from the first and third restaurants in the same commune.

Following characterization tests, 100% of strains were identified as Staphylococcus aureus and Salmonella.

The origin of the contamination in the samples was essentially due to unhygienic conditions during preparation (dirty hands, non-potable water used, lack of disinfection of materials, insufficient cooking temperature), the environment (soil, dust, etc.), and the conditions in which the food was stored, displayed and sold.

We suggest rigorous application of hygiene rules throughout food preparation, display and storage. This would significantly reduce the microbial contamination of these foods.

Finally, it should be noted that the level of hygiene and food safety for AVP remains to be discovered.

Based on our results, we note that after analysis, 80% of samples were unsatisfactory for the germs studied, such as staphylococcus. This leads us to say that the hygienic state of the preparation and sales of the second AVP would not be good for human consumption. He of strains were catalase and coagulase positive and of strains show an intermediate sensitivity to ciproflaxine. Which thus showed effective antibacterial activity on all the strains tested.

In view of our results we suggest:

❖ To the public

✓ Compliance with hygiene rules

✓ Avoid all self-medication with antibiotics, as this poses a risk that facilitates the emergence of bacterial resistance.

❖ To sellers:

  ✓ Wash hands thoroughly with soap before handling food and when leaving the toilet, and keep food ready to eat out of reach of flies, wind, etc.

  ✓ Cleaning often unsuitable kitchen utensils (knives, containers, etc.) with large quantities of water.

❖ To future researchers

  ✓ To extend this study to all health institutions in the city of Kisangani, with a view to generalizing our findings.

❖ To the authorities

  ✓ Set up a surveillance service,

  ✓ Equip our laboratories to generate good results.

BIBLIOGRAPHICAL REFERENCES

Abisa. G, 2004: Dénombrement, caractérisation et sensibilité de staphylocoque et enterobacterie isolés à partir du boudin vendu au marché central de Kisangani (RDC). Unpublished dissertation, Faculty of Science, 26 p.

AFNOR (Association Frangaise de Normalisation) (1996). Analyse Microbiologique Contrôle de la qualité des produits alimentaires (Microbiological analysis: Food Quality Control). In: Méthodes Horizontales, Volume 1. AFNOR, Paris.

Ambegebo, 2016: Microbiological quality of street foods (Case of "Liboke" sold at the Markets of Kisangani, DR Congo)

Annette, V., Brian, C., Daniel, W., Janice, G., Farland, M., Baker, R., Richard.H (2007), Vancomycin-Induced immune thrombocytopenia, New Eng. Med, 2007, 356:904-910

Barron, 2002: Evaluation de la qualité microbiologique de quelques aliments de rue dans la ville de Ouagadougou, du Burkina Faso, ah étude Rech Francoph. /santé, 1214): 369-374.

Biomerieux, 1980: Bacteriology. Paris France, 179p.

Bisa, G. 2004: Enumeration, characterization and sensitivity of staphylococcus and enterobacteria isolated from blood sausage sold at the central market of Kisangani (DRC). Unpublished dissertation, Fac. Des Sciences, 26 p.

Botchaka, S., 2013: Etude de la qualité hygiénique des aliments vendées sur la voie publique à Kisangani (Cas du Marché Limanga ya Nsé de la Commune Mangobo), 33p.

Bourlioux, P., 2000: Toxico-infection alimentaires. Service report, Danone Institute, Norway, pp. 22

Cheersbourgh, M., 2000: District laboratory practice in tropical countries, part 2n edition Cambrigole, 434p.

Delagoutte, C., 2004. Avoiding food spoilage and pathogens. Available at

Diouf F., 1992: Contribution à l'étude de la qualité hygiénique des aliments vendées sur la voie publique (AVP) dans la région de Dakar, Thesis, Université Cheikh Anta Diop de Dakar, Senegal, 77p.

Euloge S., Adjou, Yehouenou B, Codjo M. Sossou, Soumanou M. and Comlan A. de Souza, 2012: Occurrence of mycotoxins and associated mycoflora in peanut cake product (kulikuli) marketed in Benin; African Journal of Biotechnology, Vol. 11(78), 14354-14360.

Ezekiel, C. N.; Sulyok, M.; Warth, B.; Odebode, A.C. and Krska, R., 2012: Natural occurrence of mycotoxins in peanut cake from Nigeria; Journal Food control, volume 27(2) Elsevier-ISSN 0956-7135.

Etobo, 2006: Bactériologie générale, note de cours inedite, Faculte des sciences. UNIKIS

Etobo, K., 2012: Etude de l'activité antibactérienne des extraits de quelques plantes sur les souches bactériennes résistantes aux antibiotiques courants à Kisangani. Unpublished thesis, UNIKIS, Fac. Des Sciences, 142 p.

FAO, 1990: Street foods. Report of an FAO Expert Consultation, December 5-9, 1988, Yogyakarta, Indonesia, Rome.

FAO/WHO, 2007: Five keys to safer street foods, Geneva, 72p

Gentillini, 1993. Médecine tropicale, Flammarion, 4th edition, Paris, 635 pp.

Hobbs, Gilbert, 2011: Food poisoning and food hygiene

INSTITUT PASTEUR, 1983: antibiogram pasteur, determination of sensitivity to antibacterial agent 29p

Kabena, L., 2011: Qualité hygiénique des aliments vendus dans le milieu public à Kisangani (cas de milieu hospitalier), unpublished dissertation, ISTM-Kisangani, 29p.

Kalume, 2008: Dénombre et caractérisation des staphylocoques et salmonella isolés de la viande fraiche vendue au marché central de Kisangani, 34p.

Kavira, L., 2010: Hygiène alimentaire dans les restaurants de la commune de la Makiso à Kisangani (RDC), unpublished dissertation, 67p.

Kazadi, 1998: Characterization of fecal coliforms isolated from spring water in the commune of Lubunga.

Lambert, 1989: Microbiologie des aliments, Université de Louvain la Neuve, 175p.

Lambewa, Lambe, 2001. Analyse bactériologique de maser, Mémoire inédit, IFA Yangambi, Kisanhgani, 41p.

Larousse Médicale: Vainsten, JP, Larousse Médicale, 5th edition, Ed. Antoine Laron, Italy 2009, 1113p.

Like, 2013: hygienic quality of 3 restaurant menus consume at marche centrale in Kisangani. TFC,31P.

Lokangila, M., 2013: Qualité hygiénique des aliments vendées sur la publique à Kisangani, cas de marché de l'IAT (contamination des AVP par FMAT et Salmonella), p.42

Makengo, 2002: Recherche des staphylocoques à partir de quelques aliments prêts à consommer vendus au marché central de Kisangani. Unpublished dissertation, Faculty of Science.

Martial G., 2016: Appréciation de la qualité microbiologique du " Tchachanga", viande de mouton braisée vendue aux gares routières de Bohicon et de Hilla-Condji, Rapport de stage de fin de formation, Université d'Abomey-Calavi, 54p.

Mbolikoloho, 2004. Dénombrement caractérisation et sensibilité des germes isolés à partir des aliments prêts à consommer vendus sur le marché IAT de Kisangani, unpublished dissertation, UNIKIS.

Mekhloufi Omar, 2018: Search for coagulase-positive Staphylococci in restaurant foods in Algiers and molecular characterization of virulence factors.

Michael, J. ,1982: Elément de microbiologie E.C.S cham. Paris. 4061°

Mombeni, 2021: Testing for staphylococcus and salmonella on the palms of street food vendors in the commune of makiso,

Monica, C. 2000: Distriot laboratory practice in tropical countries, part 2, Combridge, low price, 434 p.

Mpiana, K., 2013: Qualité hygiénique de 3 menus consommés dans les restaurants du marché central à Kisangani (cas de lituma au poisson, fufu au poisson et riz à la viande), 24p.

Mughole, K., 2012: Contribution à l'étude de la qualité hygiénique des aliments vendées sur la voie publique (AVP) cas de marché central de Makiso, 32p.

Muyisa Soki. N., 2015. Hygienic quality of food sold on the public highway; Cas de la commune de la Makiso, Unpublished dissertation, Unikis, 36 pages.

Nzoto.,2016: Contribution à l'étude de la qualité bactériologique des aliments vendus dans la voie publique de la ville de Kisangani (RDC): cas du manioc préparé (masele).

OMS ,1988 (a). Mass catering, Regional publication. European Series Geneva

WHO, 1988 (b). Control of salmonellosis: the role of hygiene applied to animals and products technical report series, Geneva: WHO. 91p

Pene M.J., 2020: Evaluation de la qualité hygiénique des potato chips de Banane plantain et Manioc vendus publiquement dans la ville de Kisangani. Travail de Fin de cycle inédit, Faculté des Sciences, Université de Kisangani, 28p.

Reginault, J.P., 1990. Microbiologie générale, Vigot, Paris, 85p

Sabiti, 2015: Contrôle de la qualité hygiénique des viandes de porcs cuites vendus dans les voies publiques dans la ville de Kisangani. Unpublished TFC, Cepromad, 19 p.

Secke, C., 2007: Contribution à l'étude de la qualité bactériologique des aliments vendées sur la voie publique de Dakar. Thesis, Université Cheikh auto diop, Dakar/Senegal, pp. 22

Soumare, G., 1997: Contribution à l'étude de la qualité hygiénique des eaux de boissons vendues sur la voie publique de Dakar, Thèse méd, Vêt, n°1084p.

Swedi, 2021: Microbiological quality of roasted and caramelized peanuts sold in the public streets of the city of Kisangani, more precisely in the vicinity of public markets.

Trankivic, J., 2000: Glycopeptides, Encyclopédie pratique de médecine, 5-0060,506p.

Webography

www.fao.org/docrep/top of waww. memoireonline. Microbiology of street foods (2001-2012). Htm.15.14.05.05.2015

www.fao.org/docrep/top of www.memoireonline. Contribution to the study of the bacteriological quality of street foods (2000-2012).

APPENDICES

APPENDIX 1: GRAM STAINING RESULTS

Strains	**Gram staining**	**Shape**	**Color**
1	+	Hull	Blue
2	+	Hull	Blue
3	+	Hull	Blue
4	+	Hull	Blue
5	+	Hull	Blue
6	+	Hull	Blue
7	+	Hull	Blue
8	+	Hull	Blue
9	+	Hull	Blue

Sensitivity test results with three antibiotics

Strains	Amoxicillin	Ciprofloxacin	Doxycycline	Erythromycin	Gentamycin
CT1	10-7= 3(R)	35-7= 28(S)	15-7= 8(R)	31-7= 25(S)	24-7= 17(I)
CT2	10-7= 3(R)	32-7= 5(S)	17-7= 10(R)	28-7= 21(S)	27-7= 20(S)
CT3	0(R)	0(R)	0(R)	0(R)	0(R)
CT4	11-7= 4(R)	31-7= 24(S)	23-7= 16(I)	26-7= 19(S)	18-7= 11(R)
CT5	10-7= 3(R)	28-7= 21(S)	24-7= 17(I)	30-7= 23(S)	23-7= 16(R)
CT6	10-7= 3(R)	28-7= 21(S)	18-7= 11(R)	25-7= 18(I)	25-7= 18(R)
CT7	10-7= 3(R)	35-7= 28(S)	17-7= 10(R)	23-7= 16(R)	21-7= 14(R)
CT8	12-7= 5(R)	30-7= 23(S)	16-7= 9(R)	30-7= 23(S)	25-7= 18(I)
CT9	11-7= 4(R)	28-7= 21(S)	25-7= 18(I)	19-7= 1(R)2	18-7= 11(R)
CT10	11-7= 4(R)	28-7= 21(S)	25-7= 15(I)	24-7= 17(I)	23-7= 16(R)
SA1	16-7= 9(R)	30-7= 23(S)	31-7= 24(S)	28-7= 21(S)	26-7= 19(S)
SA2	10-7= 3(R)	26-7= 19(I)	33-7= 26(S)	27-7= 20(I)	26-7= 19(S)
SA3	22-7= 15(I)	30-7= 23(S)	27-7= 22(S)	32-7= 25(S)	27-7= 20(S)
SA4	11-7= 4(R)	28-7= 21(S)	22-7= 15(R)	30-7= 23(S)	30-7= 23(S)
SA5	11-7= 4(R)	30-7= 23(S)	26-7= 19(S)	23-7= 19(S)	24-7= 17(I)
SA6	13-7= 6(R)	37-7= 30(S)	19-7= 12(R)	28-7= 21(S)	31-7= 24(S)
SA7	11-7= 4(R)	30-7= 23(S)	25-7= 18(I)	24-7= 17(I)	26-7= 19(S)
SA8	11-7= 4(R)	31-7= 24(S)	27-7= 20(S)	23-7= 16(R)	25-7= 18(I)
SA9	11-7= 4(R)	34-7= 27(S)	30-7= 23(S)	30-7= 23(S)	28-7= 21(S)
SA10	10-7= 3(R)	33-7= 26(S)	25-7= 18(I)	23-7= 16(R)	30-7= 23(S)

Kolibri 360 GmbH
Ferdinand-Jühlke-Straße 7
99095 Erfurt, Deutschland
produktsicherheit@kolibri360.de

Druck:
CPI Druckdienstleistungen GmbH
im Auftrag der
Kolibri 360 GmbH
Ferdinand-Jühlke-Str. 7
99095 Erfurt